I0470626

Solar Fred's Guide to Solar Guerrilla Marketing

10 Solar Guerrilla Marketing Ideas for Residential Solar

By Tor "Solar Fred" Valenza

Copyright © 2013 Tor Valenza
All rights reserved.

ISBN-13: 978-1491274088
ISBN-10: 1491274085

DEDICATION

Solar marketing is an integral part of the solar industry, but my fellow solar marketers rarely get credit for their efforts to stand out and educate the public about the benefits of solar energy. So, this book is dedicated to all of the solar marketers, communicators, and PR people who are inspiring the world to go solar. Thank you for your courage and your creativity

Table of Contents

Introduction

This book was written for all of the residential solar installers who are making their own way in the marketing world. They may not have a marketing MBA or a large marketing budget or marketing staff. All they have is a passion for solar and providing great customer service.

Like solar guerrilla marketing, this book is a bit raw. Originally, it was intended to be an e-course with more graphics and a more web-friendly layout, plus a private social community on LinkedIn for course takers. As a result, it was also going to be more expensive.

However, to accomplish that original vision, I needed more help, more time, and resources, and in the end, much of this information was just sitting on my virtual shelf, waiting for the time to bring it all together and to focus.

And then I realized I wasn't following my own advice. I was trying to get things perfect, but solar guerrilla marketing isn't perfect, either.

So, I'm taking all the self-publishing tools that the internet has to offer, pulling it together as best I can, and getting it out there so that solar marketers can use this information, as imperfect as it may be. Otherwise, it sits on the shelf, and that helps no one.

So here it is.

First, a little Solar Fred Guide to Solar Guerrilla Marketing. Here, you'll get some basic tips, and some dos and don'ts.

Next, get ready for some serious roll-up your sleeves solar guerrilla marketing. The first four ideas in this book have never been published before. These were the ones that were meant to be part of the e-course, so their description and execution are longer and more detailed than the solar guerrilla marketing ideas that I've previously published on RenewableEnergy World.com.

After the four never-before-published ideas, you can catch up on my top solar guerrilla marketing posts that I previously published on my blog. They're not as detailed (or as expensive) as the first four ideas, but I still think they're terrific ways to get brand attention. So, if you've never read these before or never had time to execute any of them, now's your chance.

And I mean that. Please, go forth, create a budget, and use some or all of these ideas. If you need to adapt them, adapt them. Play around with them. Make them your own. That's how it should be. As a potential guerrilla marketing artist, I'm just providing you with a brush and some colors. Your execution is what will create marketing art and, I hope, great solar success.

These days, people have to be inspired to go solar, and my hope is that these ideas will provide inspiration for both you and your customers.

Solar Fred's Guide to Solar Guerrilla Marketing

- What it is
- Do's & Don'ts

What is Guerrilla Marketing? The short answer is that it's a courageous, creative, positive, and outside-of-the-box way to bring brand attention and loyalty to a product or service. For this book, that product or service is solar energy, primarily for residential and commercial customers.

If you feel that guerrilla marketing is just a "stunt" or beneath you, or that it will hurt your brand and reputation more than it will help, nothing could be farther from the truth. In fact, I would argue that more than any other time, solar needs MORE courageous, creative, positive, guerrilla-style marketing and that it can only help you to compete and succeed in an increasingly commoditized and corporate installer market.

Keeping your company's head down only allows more attention for the coal and oil industries, which are getting attention the old fashioned way: Lots of PR money, slick, fear mongering advertising, and highly paid lobbyists.

As for the larger solar companies, they'd love to do solar guerrilla marketing too, but with perhaps one or two exceptions, they can't; or if they can, they pull their punches. They pull their punches because there are too many decision makers there, and they all have to get top-down approvals to do something that's anywhere out-side of the box of traditional, boring, marketing and social media. They're afraid of what their investors and stockholders would think if they did anything beyond an expensive ad campaign or give-away.

And even when a big solar decision is made to do something "different," they hire high-priced advertising agencies that may be creative… but don't understand our solar industry and the day-to-day concerns of our customers. So, what inevitably happens is that the ad agency produces something slick that looks great and may be surprising, but falls flat. People don't get it.

You, dear solar marketer, you're more than likely a small to medium size solar company with a small marketing team. You can be nimble because you can speak and work directly with the head decision maker—especially if you're the person who's reading this.

Is guerrilla marketing always just a surprising event or stunt?

While stunts offer a lot of publicity potential, guerrilla marketing concepts can be applied to all aspects of marketing, from events, to sales, to lead generation, and customer service.

The main goal is for customers and potential customers to notice and remember you in a positive way. That imprint cements not only memory, but also brand loyalty. If the sale doesn't happen in the end, chances are they will still remember and refer you to friends, family, and colleagues for your creative efforts to delight, care, and educate them about solar.

The Qualities of Solar Guerrilla Marketing

Ideally, solar guerrilla marketing should have some or all of these qualities:

1) Solar guerrilla marketing is relatively inexpensive.

It has to be, especially for this growing solar industry. But those limits can force us to find ways to create more bang for whatever bucks are necessary to pull it off.

That being said, most solar guerrilla marketing ideas are going to have *some* cost, whether that's investing in some type of referral fee or gift to a customer, or perhaps it will only cost you time. Nevertheless, the goal is to make it as inexpensive as possible—and certainly less expensive than traditional, boring, non-interactive advertising.

2) Solar guerrilla marketing is surprising. It has a twist.

Being a solar marketing guerrilla implies that you're employing surprising and unconventional marketing methods and ideas. So, whatever marketing ideas that I suggest here or that you come up with, it should be a tactic that is uncommon in the solar industry.

Second, it should be surprisingly delightful for the customer. One of the referral ideas in my book is structured to be a completely unexpected, positive surprise that is sure to spread word of mouth about your customer service and inspire more referrals.

3) Solar guerrilla marketing shows *and* tells something.

Many believe that solar guerrilla marketing is just a stunt. It can be a "stunt," or a one-time event, but the best stunts have a specific message or goal—ideally to promote the name of your company in a positive way. So, even with a one-time event "stunt," people participating should remember your brand and its message about solar power.

4) Guerrilla marketing inspires people to act or talk to their friends and neighbors about your actions/and or customer care.

Whether it's extra-mile customer service, a surprising event, or a surprising image on your website, solar guerrilla marketing's main goal is to encourage people to either talk about you to their real and virtual social networks, and/or to contact you for a solar quote. A pretty website doesn't do that. However, a pretty website with a home page photo of puppies climbing over solar panels can. Visitors will ask themselves why you have that picture, and when they click on the photo's link, they learn that you donate $5 to the a local animal shelter for every solar quote.

5) Solar guerrilla marketing has the seed to get attention from the media.

The media is always looking for different, fun, or inspiring stories. Solar guerrilla marketing offers those opportunities. It may be just a footnote at the end of the local nightly news, but that, "And finally tonight, a local solar company has unique way of spreading the news about solar being affordable."

That's not the case for every idea, but if you do something bold, public, and creative, and then video it, there's a good chance the media will pay as much attention as onlookers.

6) Solar guerrilla marketing is non-violent and legal.

Whatever we do, it always has to be legal and safe. This really doesn't need much explanation here. While guerrilla marketers should be creative and take creative risks, ideas should not endanger anyone, including employees, and they should not stretch or break any laws or regulations.

All the ideas in this book are designed to be legal and safe, and I expect that when any solar company applies these ideas to their own businesses, that these ideas will remain safe and legal, as intended.

That's not my lawyer saying that; that's me, Solar Fred. Please be safe and do not break any laws, no matter how passionate you are about solar.

7) Solar guerrilla marketing isn't perfect.

If it's guerrilla marketing, there's no real rulebook. That doesn't mean you should ignore everything I've just written above, but it does mean that you may not execute your own ideas or my ideas perfectly. Perfection is for machines, and even they malfunction. Solar guerrilla marketing is more like an art, and as such, you may have an outline here or in your own mind, but the final product may come out unexpectedly better ...or worse.

Whether you use the ideas in this book or are inspired to create your own solar guerrilla marketing ideas, be like Nike™, and "Just do it!™" Don't wait to get all of your ducks lined up. Just make a plan, believe in its purpose, and start executing it with whatever resources you have.

Because it is fun or original, you will get attention, and you may even get better results than you expected...or not. So, yes, there's some risk here, whether it's time or financial. But only doing your traditional marketing methods will also gain you nothing. No extra brand recognition, no extra leads, and no extra referrals.

In summary, be **bold** for solar. Stand out and educate.

Happy Birthday, Solar System!

A fun way to cement brand relationships and referrals for years to come.

THE SOLAR MARKETING CHALLENGE

We all know how happy customers can become a referral engine, recommending you to friends, family, and colleagues. The challenge is to continually engage these customers and remind them about their solar savings, as well as what a great installer you are.

Often, it's not only the customers that forget you, but you may also forget them once their systems are all up and running. As solar pros are well aware, solar systems can last 25 years or more. So, wouldn't it be great to have a systematic way to intermittingly and automatically engage with your past customers once a year and have them actually notice...and smile?

The solar guerrilla marketing solution: Celebrate customer anniversaries every year with birthday cake, champagne, or some other delightful annual surprise.

14

WHY CELEBRATE SOLAR INSTALLATION ANNIVERSARIES WITH SURPRISE GIFTS

As noted above, it's difficult to engage with past solar customers without being predatory and annoying. Calling can be invasive, and emails and direct mail can be easily tossed away.

Nevertheless, genuine customer service and care is invaluable. People smile when they receive something out of the blue for no other reason, except to say "Thanks for your solar business." Having beyond-the-call customer service and care will keep you on customers' minds at least once a year. And if you give them something delightful, memorable, and shareable, they may even talk about it to their friends and family, co-workers, not to mention review websites, like Yelp!

As we'll see, not only can an annual token gift bring back fond memories, it's also a way to convey more information about new services, such as energy efficiency, or referral programs.

ACTION PLAN AND BEST PRACTICES FOR CELEBRATING CUSTOMER ANNIVERSARIES WITH GIFTS

Preparation

The first thing you'll need to do is to create a calendar for all past customer (and future) solar installations. This can be a physical paper calendar, but it's probably best to have an online electronic one using Google apps or some other sharable calendar software.

In terms of customer data on the calendar, it should include:
- Names of heads of household
- Children's names, if known
- Address
- Email
- Date of solar system going online
- System size

The first year of any solar system is important for your business, as well as for the homeowner. The solar system is new for them, and they're going to be concerned about it performing well.

Consequently, I recommend that during the first year after installation, that the customer be contacted three times during the first year.

Using your computer's calendar application, set a reminder for the following:

Contact#1: A phone call one month after installation. Even if you're sure the install is routine and perfect, have the sales person or customer service department call the new solar owner and check to see that all is working well and if they have any questions. If you leave a message on a voice mail, be sure to say that no call back is necessary, but that you're just checking in.

Contact #2: An automated email after 6 months. Create an automated email through an email program, like MailChimp (**www.MailChimp.com**), which says that the solar system is now six months old. Congratulations. Include a reminder to clean the panels. It may also be a good time to remind them that solar systems perform best during the spring and summer months, and have lower performance during the fall and winter months.

Contact #3: Happy Birthday! Your solar system is a year old. We'll get into the specifics of this below, but here is when you will celebrate the one-year anniversary of each customer's solar system and send a small gift.

After the first year, you can repeat the anniversary gift for the next 10 years....or longer.

Setting Up The Anniversary Gift

You can be as creative and as lavish or economical as you wish. You know your town, and you know what people like and value. Nevertheless, I would shoot for a gift that has a net cost to you of between $10 and $20 per customer. Keep in mind that this investment in customer care and delight can pay off handsomely with just one referral, and perhaps many more.

What type of anniversary gift should you choose?

Do something that is easy to deliver and that is shareable; make it special. For example:

- **Make a deal with a local bakery that can create solar-themed cupcakes, cookies, or a small cake.** I like this gift the best because dessert usually brings a smile to people's faces. Don't worry about individual allergies, such as nuts, gluten, or chocolate. That can be seen as private information, and no one's forcing them to eat it. It's the thought that counts. If the family has any of these issues, they will probably be able to give the cake to a neighbor or take it to the office. "Where did you get that from? It came from my solar company. My solar system is a year old." That will inspire solar conversations about your company.
- **Make a deal with a local pizza place.** Similarly, you could mail a gift certificate for a cake or a pizza. If you do this, place the gift certificate in a small gift box so that it's noticed and not trashed with junk mail. Include a note about the anniversary.
- **Send champagne, wine, or beer from a solar-powered brewery or vineyard.** Once again, make a volume deal with a local liquor store or online retailer. Remember that local laws about sending wine and spirits by mail vary from state to state, so be sure to follow those guidelines.

- **Send two movie tickets from a local theater.** Once again, place tickets in a gift box so that the package is noticed.

The above are general suggestions, but you may have more ideas for partnering with local vendors. Whatever gift you send, include a standard note about the solar system being (another) year old, and congratulate the owner.

In addition to a gift certificate, you can also include separate sheets of marketing materials in the gift box, such as, "Know someone who could go solar? Give them this card, and they can save x dollars off their install." You could also include sheets of other offers, such as monitoring, maintenance contracts, or extended inverter warranties. You could provide the same materials and brochures to your bakery partner.

Could you just send a personal postcard or email every year? Sure, you could, but would it have the same memorable impact as a small gift? I don't think so. Emails and postcards can be deleted or trashed immediately. It's less likely that an anniversary gift will be ignored.

Continue Annual Anniversary Presents for at Least 10 Years

By using an electronic calendar with a reminder system, you can continue this anniversary gift giving for as long as you want, and it doesn't have to be the same gift every year.

Surprise is part of the "wow" factor of receiving this anniversary gift. Customers will be pleasantly surprised to receive a cake, pie, or other small gift from you every year. And if you vary what you send every year, that can add even more to the surprise.

Additional Partner Marketing

In addition to creating marketing partnerships with local, native marketing partners, you can also create a referral program with these vendors and pay them a sales-based fee for every closed sale of a solar system that uses a special referral code.

Partners will then be more open to displaying brochures in their store, potentially creating even more referral prospects through this anniversary initiative.

If you have a customer referral program that pays a fee for every closed sale, include a sheet for this too.

Your annual small-token generosity and surprise gift will remind your customers about their solar system and who installed it.

SUMMARY

- Celebrating sales anniversaries reminds customers about your service and provides multi-year opportunities for referrals and increased word-of-mouth.
- Working with a calendar and local vendor partners, you can provide an annual solar installation birthday gift that surprises and delights customers, inspiring them to talk about you to their friends, families, and social media networks.
- Your annual gift can also establish new marketing partners that can create referrals and brand recognition.

Your Mobile Solar Guerrilla Marketing Store

Are you really tied to a garage or a warehouse or a low traffic part of town? For a small investment, be where your customers are.

THE SOLAR MARKETING CHALLENGE

Solar companies don't often get a lot of walk-in street traffic because they're often based in warehouse or commercial districts that consumers and shoppers rarely frequent.

And yet, consumers are curious about solar and saving money, but they aren't motivated to get educated about it. So, the challenge for solar companies is finding the opportunities to educate them and to get their attention. You have to be in the right place and the right time. And to find that place, you have to consider your customer demographics: homeowners with a high enough credit rating for a solar lease, home loan, or cash sale.

The solar guerrilla marketing solution: Invest in creating a mobile solar storefront that brings you to where your customers are.

WHY DEVELOP A MOBILE SOLAR STOREFRONT OR SHOPPING MALL KIOSK?

The idea here is to be where you customers are, but without paying regular commercial retail rent. And where are your customers? They're at the high end indoor and outdoor malls, walking the aisles and shopping. They're at farmers markets, or they're outside the busy office complex where many walk outside for coffee breaks, lunch, and meetings.

SolarCity and Sungevity are already doing this with Home Depot and Lowes home improvement stores. They've set up little kiosks staffed by one employee who can attract customers, hand out brochures, capture lead information, and give quick solar estimates via laptop. They're also increasing their brand recognition, even if nobody stops to pick up a brochure.

Any solar company can do the same thing at another hardware store, mall, or with a solar push cart. It takes a little investment, but when effectively executed, the leads and sales generated can more than pay for that initial investment.

Action Plan and Best Practices for Mobile Solar Storefronts or Kiosks

Location

First, think about where you might set up your mobile kiosk or shop. That will determine whether you need to build or adapt a cart on wheels, or whether you can set up shop and adapt to a pre-built mall kiosk.

8 Examples of Mobile Locations:

1. **A popular outdoor mall that takes advantages of sun.** With an outdoor kiosk, you can have an active AC solar panel for demo purposes, as well as for things like having a solar cell phone charging station.

2. **A popular indoor mall with a high traffic area.** While you may not be able to have an active solar panel charging station, you can still demo and sell small solar toys to attract attention. Some malls also have skylights, and you may be able to position your cart there for active demonstrations.

3. **A Weekly Farmer's Market.** Farmer's market customers are already interested in sustainability. Set up a solar charging station and become a regular exhibitor.

4. **Having a licensed pushcart outside a popular coffee shop frequented by business people.** Once again, here you can design an active AC solar panel for demo purposes, as well as a cell phone charging station.

5. **Having a licensed pushcart outside a large office complex.** If people can sell hot dogs and pretzels from a cart, you can have a solar cell phone charging station and do solar estimates.

6. **Having a licensed pushcart outside a courthouse.** Lawyers, judges, clients, and staff often have homes, and they often take a break outside the courthouse.

7. **Having a licensed pushcart outside a zoo, aquarium, or other local family venue.** Families go to zoos, and families often have homes with large electric bills. With solar toys to attract attention, or a solar cell phone charging station, you can strike up a conversation with Mom and Dad about saving money with solar.

8. **Having a licensed pushcart in a family friendly public park.** Once again, families go to parks. Find the right location to set up shop near a food venue. Then set up your solar cell phone charger and other AC panel solar demos, and start a casual conversation about how affordable solar is today.

Wherever you decide to set up a mobile solar kiosk, just be sure that it's an area where your potential customers are; i.e., a homeowner with decent credit rating.

Also, before signing any long-term lease with a mall or farmer's market, ask the mall leasing agent about customer demographics. They should include a high percentage of homeowners. If not or they don't have that information, find another location or negotiate a short-term lease to experiment there.

Pushcart and Kiosk Design Resources

www.concessionequipmentdepot.com

www.allstarcarts.com/index.html

www.kiosksinc.com

www.turnkeykiosks.com/kiosk/custom-kiosks

www.lvmannequins.com/custom-kiosks.html

Kiosk/Pushcart Design Best Practices

Once again, these best practices will depend on your location needs and any mall design restrictions or city ordinances if you are going to create a mobile kiosk or pushcart platform.

Many malls have their own design firms that force vendors to use these services and conform to certain design elements for aesthetic consistency. If you sign a lease with a mall, be sure to describe to the rental agent your goals and the design elements that you'd like to incorporate; i.e., solar panels on the roof of the kiosk, or as a shade element, etc.

For pushcart kiosks that will be removed and stored on a daily basis:

- **Be sure that the design can fit inside a van or a pick up truck.** If you are creating the pushcart yourself without a professional pushcart designer, DO NOT design with towable trailer hitch. Different states and the Feds have vehicle codes for this, which is a can of worms to be avoided. Instead, design your pushcart to be small enough that it can be rolled into the bed of a pickup truck. If needed, make it modular so that you can easily assemble every time you set it out. If you need it to be towed, buy a separate trailer (already up to codes) and strap your kiosk to that. If using a professional kiosk designer, they will be able to create a mobile trailer that meets vehicle codes. Also, be sure that they design loops or holes that can be used for pickup truck straps with bungee cords or ropes.

- **Expect to pay anywhere from $1000 to $5000 for a professionally designed custom mobile pushcart.** This may seem like a lot, but it's a great investment that will pay off for years to come if you deploy your solar pushcart regularly.
- **Design your pushcart to have electric power and Internet.** This will mean having a solar panel with a battery back up, as well a mobile hotspot.
- **Design your pushcart to be *active*.** If you have a mobile pushcart, you're going to be out in the sun, a perfect environment for solar demonstrations. I've already suggested incorporating a solar charging station into your design. Be sure to purchase the USB charging cords for major smart phone brands, including iPhone(s) Samsung, HTC, Windows, and Blackberry. Then put a sign on the cart or standing nearby that says something like "Charge you smart phone now with free solar power!" Other active ideas:

  - **Have coffee maker plugged into the demo solar system.** Give away "free solar brewed coffee." (Be sure your cart complies with all local food ordinances.)
  - **Have a mini-refrigerator built in.** Give away "Free solar cooled water (or Coke™ or lemonade) today. (Be sure your cart complies with all local kiosk/pushcart food ordinances.)

24

- o **Sell solar toys.** There's no reason why your cart can't also sell solar-related items, as well as get solar quotes. So, stock your cart with solar electronics and toys. Sell those, but also offer to give away an inexpensive solar toy for anyone who signs up for a quote.
- o **Run a fan off solar.** If you're in a hot area, create a low power solar air conditioning system with fan and a block of ice. Post a sign that says, "Cool off yourself—and your home—with free energy from the sun!"
- o **Have a laptop/iPad solar quote system.** Your kiosk could be one big electronic solar quote kiosk, where visitors input their average monthly electric bill and zip code and get a quote, or where you have your own estimation system from one of the many quote software companies, such as Ongrid Solar or Clean Power Finance or any leasing company that you work with. Whatever computer hardware you use, be sure to design the system to be tied down and secure so that someone can't walk away with the hardware. Your sign should read something along the lines of "Get a Free Solar Estimate Now for Your Home or Business."
- o **Have an ongoing solar information class.** Create a 5-10 minute PowerPoint video presentation, or 100% verbal class that you can run every half hour. Have a sign that says "Solar Energy Information Classes Start Every 30 Minutes."

- **Include a back up battery system.** Not only will it help with your charging, it can also be used as an analogy for explaining net metering.
- **Have the micro-inverter visible.** Whether you use micro inverters or strings in your installs, make sure your live demo panel has a visible inverter. This will also help you explain solar to prospects that come to your cart.

- **Design slots for brochures, clipboards, and other FAQ information.** You don't want to have people waiting to talk to you when you're busy with a prospect. If your kiosk has only one employee, make sure you have clipboards with a contact sheet and pens to silently hand to people who may be waiting to speak to you. Also, have wind-proof slots where people can easily grab a brochure, an FAQ sheet, or a How Solar Works diagram.
- **If possible, design your pushcart solar system be a shade/canopy for the pushcart.** Solar shades can look very pretty, especially with Sanyo/Panasonic HIT panels that are made to collect light on both sides. You can also purposely position the cart to face south for further live customer education. Perhaps more importantly, a shade can help waiting prospects to keep cool as they talk to you.
- **Don't forget your company name/brand and logo and website.** Be sure this is prominently displayed on all four sides of your booth. Even if no one stops, you'll get brand recognition for the future. It's also important to work your web address into that signage. In terms of contact info, website URLs are even more important than phone numbers today.

For static kiosks in a mall that stay in the same place

If you are going to be locating your solar kiosk in a mall, indoor or outdoor, many of these design elements may be moot, as the mall will have some restrictions on what can or can't be incorporated into your kiosk. If possible, ask mall leasing companies if you can incorporate any of the elements mentioned above for the mobile solar pushcarts. Nevertheless, there are some design elements that shouldn't be affected by whether or not you incorporate a live solar system.

As mentioned earlier, your goal is to become a remote solar information and a solar estimate booth for lead generation. For this purpose alone, you can incorporate into any kiosk, indoor our outdoor, with little design controversy:

- Acrylic brochure holders or built-in slots for brochures, FAQ sheets, etc
- A computer monitor with informative videos running. This may include pictures of installs, video testimonials, and more.
- Laptops or iPads that are able to generate quotes with solar estimation software.
- A cellphone charging station.
- Solar toys and gadgets that you can sell.
- Solar information class (Be sure "students" don't block mall aisles.)
- Poster images of beautiful home and commercial installations that you've done
- Signs on every side with your company name and logo, and website.
- Secure. Any valuables, such as electronics or sales props, will need to be able to be stowed away and secure from theft.

BEST PRACTICES FOR MANAGING YOUR PUSHCART OR KIOSK

Staffing. Your new solar kiosk has one real purpose: To generate leads for your installation business. So whether you sell solar toys or give away free solar coffee, your booth staff must know solar FAQs, including installation, policies, and incentives, and be able to encourage people to sign up for a solar quote.

Start with one full time staff member working 5 days, and a second part time staffer for the remaining two days. If that person is overwhelmed, it may be necessary to hire people for more hours. This will add to your costs, but this is good news, as it means you're generating many solar quotes.

Dress: Like any sales person, your kiosk staff should wear a polo shirt or button down shirt with your logo and be well-groomed.

Staff Actions: For your investment to pay off, your kiosk staff member will need to be active as much as possible. When not educating customers or signing them up for a free home solar quote, they should interact with customers passing by with a call to action and holding a brochure.

As the staff member says the call to action, he or she should offer a brochure to anyone passing by, but especially people with children who are likely to own a home.

Depending on your both design, sample calls-to-action may include:

1. Would you like to learn about saving money on your electric bills with solar?
2. Have a free solar powered cup of coffee/Coke™/bottle of water and learn about solar.
3. Need a quick cellphone charge? Feel free to use our free solar charger.
4. Did you know that you can now go solar with no-money down?
5. Take two minutes for a free solar estimate and find out how much you can save on your home's electric bill. Or just "Find out how much you can save with solar."
6. If you think solar is too expensive, don't guess. Let me give you a free estimate right now.
7. Finally, solar is affordable. Find out how much it costs for your home.
8. Curious about how much solar costs? Get an estimate right now.
9. If you've ever wondered about how much solar costs, let me give you a free estimate in 5 minutes.

10. Sign up for a free solar energy class. A new one starts every 30 minutes.
11. Sign up for a free solar quote and get a free [product give away].
12. We've got a $x solar installation discount today. Let me give you a free estimate.
13. If your electric bill is $100 or more, solar can cut your bills without any upfront costs (with a solar lease).
14. We have a new no-upfront cost to finance solar.
15. Did you know that you don't need batteries to install solar today?
16. (for SREC states) Did you know that new solar incentives could pay you x dollars a year to go solar?
17. Did you know there's a federal solar incentive that can knock 30% off the price of going solar?
18. Ever wondered how a solar system is attached to your roof without leaks?
19. Want to learn how you can lower your electric bill by %?
20. Did you know that homes with solar systems appreciate 17% more and 20% faster than non-solar homes? (Offer a handout that summarizes this study: **http://ow.ly/nATym**.

Staff Incentives and Kiosk Performance

Like all sales staff, it's important to incentivize performance. So, offer a local commiserate base salary plus a bonus that is payable for every home visit generated by the kiosk. You'll need to keep track of these in an Excel sheet or some other lead generation software, such as SalesForce.

Be sure to pay the bonus monthly and to track how many convert into sales. Not only will this assess your staff person, but also the success of the leads being generated by having a remote lead generating kiosk.

If performance for your kiosk is not generating an ROI by three months, it could be due to many factors:

- **You may have a poorly designed kiosk.** Ask friends and colleagues about the design and whether it's attractive or looks cheap and shoddy. Ask them to be specific about any poor functionality or design elements. Consider a redesign with a new kiosk/pushcart designer. Perhaps staff sitting in a chair or on a stool is uninviting, so perhaps eliminate that.
- **You may not have a great sales person(s).** Try new people at the booth and see if your leads don't improve. As noted, kiosk personnel can't be flies on the wall. They must be actively engaging with potential clients with calls to action. If leads are coming in slowly, observe their performance through a colleague or friend that does not know the kiosk employee. If he or she is active and courteous, then there's another problem.
- **You may not have enough going on to attract people to the booth.** People may not need to have their cell phones charged. Nevertheless, for guerrilla marketing purposes, it's important to stand out in *some* way. If one of the above suggested ways to stand out is not working, you may have an even better solution for your community. You know your community best. If you're in NY or Boston and giving away free Dunkin Donuts will make people stop and listen, make that investment and see what happens. The hook may vary by region. Be unique and fun if possible, but always remain courteous and family friendly. (In other words, don't emulate Hooters.)
- **You may have the wrong location.** Hopefully, you did sufficient research before signing a long-term contract with a mall location. If you have a mobile kiosk, change your location and see if that improves something. Some days may be better than others, such as weekends for parks and zoos. If you have a static booth, that's a bigger challenge. It's not easy to start over again at another location, but it may be worth it in the long run. If your lease is not month-to-month, stick

30

with it and try new employees and new engagement tactics.

SUMMARY

- Creating a mobile or satellite solar lead generation kiosk is a great way to get to a Main St location and be where your customers are.
- Thoroughly research your location and the activities you will employ at your booth. Design an attractive booth and hire an engaging staff person who is knowledgeable about solar and sales. Be sure to have plenty of useful literature to offer prospects.
- When successfully executed, your kiosk can provide brand recognition, residual income, customer education, and most importantly, targeted lead generation.

Solar Guerrilla Marketing FAQs

Making standard solar FAQs come to life.

THE SOLAR MARKETING CHALLENGE

As much as humans enjoy and continue to read, the truth is that we are increasingly living in a short-attention span, visual world, especially when it comes to the Internet.

A page of solar Frequently Asked Questions (FAQs) is not only useful to your solar customers and a time saver, it's also something that they can share with other family members or associates.

Many solar companies don't even provide FAQs on their website. If you're one of those companies, this solar marketing tip will give you a list of the most important ones. And if you do have solar FAQs, this idea will take them one step further.

The solar guerrilla marketing idea is not just providing useful text-based solar FAQs, but *video solar FAQs*. Not only will this be more visual and easier to digest, video solar FAQs can be repurposed in many, many ways.

WHY VIDEO FAQs

In today's bling-bling and exciting Internet world, consumers are easily distracted. So despite your good intentions to provide useful but brief frequently asked questions (FAQs), people may quickly move on, intimidated by all the text, even when it's brief.

Brief solar FAQ videos are an answer to the intimidating wall of text. The old saying that "a picture is worth 1000 words" is even more meaningful for video.

The advantages of a solar FAQ video:

- Video can show your solar company's voice and personality.
- Video is visual, so you can not only talk, but also *show* examples.
- Video is shareable. Once you record and upload your videos to YouTube, these videos can easily be shared on Facebook, Twitter, and other social networks…and be shared over and over again.
- Video can be watched and shown to family members on iPhones, iPads, and other mobile devices. Mobile devices are increasingly becoming important for consumers who want information.

A while back I wrote these quirky solar FAQs on SolarPowerRocks.com (**http://ow.ly/nATBc**), a solar blog and referral website for One Block off the Grid (a.k.a. 1bog). But I have to give credit for the video idea to my colleague Dave Llorens. Dave took the FAQs to the next step, and in this guerrilla marketing book, we'll be going even farther.

What Dave did was very simple and basic: He just set up a camera in front of his desk and just started shooting video solar FAQs, one after the other. Check out a few here: **http://ow.ly/nATCI**

You can imitate his method, but it's a little too "talking head," don't you think? Plus, the execution could be better. It's a little unrehearsed and low energy.

You can do better. Here's how.

ACTION PLAN AND BEST PRACTICES FOR CREATING A SOLAR VIDEO FAQ

Talking Head Video FAQ— The Basics.

Dave's talking head video FAQ is good, but it could be improved. Follow these guidelines if you choose to do this simple talking head method:

1) Choose a location that is light and quiet. If you need extra lighting, bring it into the room. You may want to film outside, but if you do, remember that street noise may distract, so be sure to put yourself in a quiet place.

2) Select your best salesman. There's a reason your best sales person is successful. They're probably personable, attractive, and can speak easily and confidently. All of these qualities are great for video too. Have them dress as they normally would for a sales call. If they usually wear a company polo shirt or button-down shirt with a logo, even better.

3) Rehearse. Even if you or your designated sales person has answered these questions over and over again for the past 5 years, you should rehearse at least a few questions in front of colleagues, or at least a camera. People who are normally extroverted in person can sometimes freeze up in front of a camera. Rehearsing will get them used to the camera, plus reveal fidgeting or doing something distracting with your body or voice.

4) Be confident and have energy. Whoever the speaker is, make sure he or she has energy, but isn't bouncing off the wall. The trap with knowing these answers backwards and forwards is that you may come off bored spewing the answers again. An acting tip is to imagine you're speaking to your favorite person in the world, someone who cares about you and your success. This stage fright technique will not only give you confidence, but also give you a grounded feeling of being yourself, because you'll know this person loves everything you do and how you look on camera.

5) Don't write a script. You should have all of your questions (suggested FAQs are below), but don't write a script. This should be a conversational video, as if you're talking to a sales prospect. When people script things out, they're often stilted and thinking of the words they have to say instead of the content. You know these answers, so you don't need a script anyway. Instead, just have some notes in your pocket before every question. Check them out before recording each question, so that you'll be reminded about the points you want to make for each question.

6) Be brief. You'll probably be editing these video FAQs, but think in terms of one minute answers, or up to two minutes, max. The problem with talking heads is that they can quickly get boring, no matter how dynamic the person is.

7) Frame the video above the waist, perhaps midway up the chest. You don't need to show your entire body. Frame the video just above the waste or halfway above the chest. You might have an initial wide shot that shows where you're sitting in your office, but after the segment starts, quickly cut in closer to the chest shot. You also don't want to be too close, literally turning you into a talking head. Think of a news anchor behind a desk.

8) Don't have a distracting back ground. Ideally, you should be standing in front of a beautiful solar home. If you stand in front of something active or moving behind you, viewers will get distracted.

9) Introduce yourself for every question. "Hi, I'm Joe Smith from UnThink Solar, and today we're answering the question, 'How much does a solar system cost?'" This may be a series of videos, but people may not see the series in order, so it's good to introduce the speaker every time, as well as the question.

10) Have a title for each question. With your video editing software, be sure to include 10 or so seconds of a written title that shows the question that you're answering for this segment. Make sure the title is positioned below the person's chest. In fact, you may want to include this title during the initial wide shot, and then have it fade out when you cut in closer.

11) Close every question with a call to action and contact info. At the end of your answer, always ask for a solar quote in some fashion. "So that's what net metering is. If you have more specific questions about your home or would like a free quote, email me at joe@solarcompany.com or give us a call at 555-555-555."

Going Beyond the Video Talking Head

The above talking head solar FAQ will work, but it has its limits. It's not very visual, and you're chained to a desk or perhaps a single static location. However, there are certain solar FAQs that lend themselves to more visuals.

For example, "How are solar panels attached to my roof?" With this answer you could take a video of racking being attached to a home roof and narrate how flashing and sealants protect the homeowner from leaks.

Let's take my favorite Top 10 Solar FAQs for home owners and see how we can make them visual.

1) How much does a solar system cost?

This is absolutely the most frequently asked solar FAQ on the planet. The short answer is always "It depends." So here's a great way to edit in graphic representations of the variables with a voice over.

- You could show an electric bill and explain how it depends on how much of your electric bill you want to offset.
- You could mention and show different types of roofs and why it matters, as well as mention how the amount of sunlight and roof orientation can affect cost.
- You could also just have a graphic of "Lease or Buy?" That being said, don't frustrate the viewer. They're looking for a ballpark figure. If you have a solar calculator on your website, direct them to that and show a photo of the calculator or the link to the calculator. If you don't have a calculator, then give them a range— after incentives— for small house to large house. "Or, it could be as little as 0 upfront if you lease."

At the end of the segment, make sure you say something like, "Because every home is different, it's best to get a free quote and find out how much solar actually costs for your home. Email us or give us a call at..."

2) Do I need batteries?

Here, you could answer this question by standing in front of a home's bidirectional meter, and explain how net metering is like a virtual battery. If you provide hybrid back up systems, you could show that too.

3) What's a solar lease or solar PPA?

Couch this around financing solar, and briefly explain both the advantages and disadvantages of leasing/ppa versus buying. Visually, you can edit in graphs that you may typically use in your home presentations.

4) What happens when it's cloudy or raining, or during the night?

Here you could also stand in front of a meter and show how rain or shine, day or night, the home's electric power is always backed up by the utility.

5) What's an SREC?

Few might actually ask this question, even in SREC states. But if they're worth anything in your state, you should have the question and answer. Any financial benefit is an enticement to get a free quote. Visually, you could edit in a graph of how SREC prices have risen...and fallen over the last few years. Or, you could just show a customer receiving an SREC check in the mail.

6) Do I need to replace my roof to go solar?

Give your standard answer. I've heard many ranges, so I won't discuss here. Be sure to point out that you can refer them to a quality roofer that will give them a discount if one is needed. Visually, you can give this answer on top of a roof or in front of a demo roof, if you have one in your show room.

7) How does solar work?

For solar PV, show video or static photos of the various parts of one of your completed installations. Over these images, explain how sunlight hits the panel, generating electricity, etc. Be sure to mention that there are no moving parts, so very little maintenance. For solar hot water, a diagram for a closed loop system would be helpful.

8) Do you provide any type of financing?

This is another opportunity to mention your leasing program or any loan programs you have with a local bank or solar panel manufacturer. It may also be good to mention that the combined loan/leasing/ppa payment and new utility bill can be x% to x% lower than the old utility bill. Visually, you could show a bill being cut in half, or you could have a voice over and show a sales person talking to a customer and explaining financing.

9) How long does it take to install solar on my home?

Be honest here about your install times. Also, mention typical wait times for your city to approve. Transparency is best. Visually, you could just edit together snapshots of the installation process, ending with a beautiful shot of a solar home. This might also be an opportunity to explain the answer with a voice over and show a time-lapse video of one of your projects. Time-lapse videos are very cool and visual.

10) What rebates and incentives are available?

Mention any local rebates, SRECS, and the 30 percent ITC, and that because every home is different, it's best to get a free accurate quote "to see how much you can save." As you explain, you could visually show how your company does all of the paperwork for the rebate. Or you could show a video of a customer getting a rebate or SREC check in the mail. Be sure to end with a shot of the solar home.

With all of the above, remember to include a call to action at the end, such as "Have more questions? Get a free solar quote and find out if solar is right for you and your home."

Sharing Your Solar FAQs.

The other powerful aspect of a video FAQ is how shareable they are. Once you upload this video to your YouTube channel, you're able to spread your useful information—and your brand—far and wide, with the potential of others sharing your videos too.

Tips for sharing:

1) Place all of your solar Video FAQs on a single web page on your website.

And don't hide this web page in your blog or somewhere under "About us." Place it somewhere on the home page next to your Contact Us with "Video FAQs" button, or on the menu bar. Watch your Google Analytics and see how long people stay on that page. The more time they spend, the more you know you've created some successful videos.

2) Write a blog post that includes each video FAQ. For those who read my UnThink Solar blog, you know how important I think blogging is to Internet SEO and gaining authority. So, repurpose each video FAQ into a blog post that digs a little deeper into the subject than your 2 minute video. Of course, share each blog post on Facebook, Twitter, Stumbleupon, and any other social network you use.

3) Do a Twitter campaign. Once you have all of your video FAQs done, do 10-hour Twitter campaign. That is, once an hour, tweet each Solar FAQ with the question and link to the video. You'll be amazed how many RTs you get. People love to share useful info. Using social networking tools like Hootsuite (**www.hootsuite.com**), you can schedule these in advance.

4) Upload the videos to Facebook. Unlike Twitter, Facebook users don't want to be overwhelmed with content from businesses. So, rather than 1 video per hour, upload each video once a day for 10 business days, or for however many videos you've created. Announce that you'll be doing a daily solar FAQ for the next 10 days, and ask followers to share these videos with their friends. Make sure to create a specific Solar FAQ "album" on Facebook. That way, you can send that album link to prospects, allowing them to view the FAQs on Facebook; and if they like what they see, perhaps they'll share the videos with Facebook friends and family as they go through them.

5) Create a drip email program for prospects who request more information. When you take a prospects name over the phone or email, ask if they'd like to receive a Solar FAQ email that will educate them about solar power and the installation process. If they say yes, add their name to an e-newsletter drip campaign, such as one provided by Constant Contact or MailChimp (**www.MailChimp.com**).

I described the drip campaign process in a recent blog post. (See "Drip, Drip, Drip: How to Effortlessly Educate Solar Customers with a Few Mouse Clicks" in this book.) Please refer to it for a how-to video and more information.

6) Have them running in the background at home shows. Want to attract more attention to your home show booth? Have a laptop and monitor set up where people can congregate and watch all of your solar FAQs, one after another. Once people are stopped and listening to your solar FAQs, find the appropriate moment to step up and ask if them if they have questions about the video, or would they like these videos emailed to them (via the drip email campaign.) It's a great, visual way to capture show leads.

SUMMARY

- Solar video FAQs are an excellent way to provide and share useful information with prospects.
- You can create a simple "talking-head" FAQ or a more visual and active version. Either method is valuable and more effective than text based solar FAQs.
- Solar video FAQs are shareable. Be sure to not let this resource to sit on your website and YouTube. Share them individually on social networks, as well as through e-mail drip campaigns.

Celebrate Every Sale

After-the-sale solar referral parties are great. But are you really celebrating them?

THE SOLAR MARKETING CHALLENGE

You've made a solar sale, and that's wonderful, but if you allow this new customer to forget about you, your great installation, and your dedication to building solar, you've lost many more sales opportunities from referrals from these happy, but forgetful customers.

The solar guerrilla marketing solution: Celebrate every sale in a unique way that not only strengthens your new customer relationship, but also builds a memorable and lasting impression that is shareable and that can be spread through your customer's social networks.

WHY CELEBRATE EVERY SOLAR SALE?

There are several reasons why celebrating referrals is important, and it's not just related to generating referrals.

- **Celebrating ever sale motivates your sales person.** The sales process can be long and hard. With a "celebrate every sale" attitude, you're not only helping your customers to enjoy their new solar system, but you're also congratulating yourself or your sales staff on completing this long sales journey.

- **Celebrating every sale helps you bond with customers.** Large purchases make people nervous. Even with the best sales relationship, customers can often have doubts: "Could I have gotten this cheaper?" "Will something go wrong?" "What happens if new technology comes along?" Celebrating the sale gives more confidence to the customer that these concerns aren't applicable to your company and that they made the right decision.
- **Most importantly, celebrating the sale is a powerful lead generating opportunity.** It's never a mistake to delight and to be generous to your customers. Not only will they be more forgiving of any installation snafus, celebrating each sale will inspire them to refer you to their social networks and promote lasting brand relationships that can continually generate leads

ACTION PLAN, BEST PRACTICES, AND SAMPLE SUGGESTIONS FOR CELEBRATING EACH CUSTOMER SALE AND BUILDING REFERRAL OPPORTUNITIES

The sky's the limit for celebrating each sale. I'll be giving a few suggestions here, but the goal is for you to celebrate your sales in a way that is unique and memorable for your company and your company's personality.

That is, celebrating a sale in Texas should be very different than celebrating a sale in Boston. In fact, celebrating a sale for a solar company in one city should be different than another company celebration technique in the same city.

To help you think about how you can celebrate a sale in your company's unique way, ask yourself these questions:

Who are my customers? How old are they, and what's their home life like? Do most have kids or are they retired couples? What would bring a smile to their faces? Because this answer may be different for different demographics, it's a great idea to celebrate the sale in different ways for two or three customer demographics.

What is the personality of my region? Once again, if you're going to celebrate a sale with a catered party or a gift certificate to some place, be genuine to your community's interests. Do people really like BBQ where your solar company is based, or would they rather have a gourmet Pizza Party?

Similarly, do people often go to community sporting events, like high school football games? Perhaps there's a celebration gift that would be applicable to that, such as a pair of folding chairs or team T-shirts.

What is my company's personality? Have you ever thought about what your company stands for? Do you truly think about your mission statement and apply it to everything you do in the solar business? If so, that's great. If not, perhaps now is a good time to think about that again and apply that awareness to your celebration theme.

If your mission statement is to give 10,000 homes solar energy independence, one roof at a time, then perhaps think of ways to celebrate with a patriotic or self-reliance theme. If your mission is to provide the best customer care in the solar business, how can you show that "care" with a celebration gift? Could you offer a free massage service that literally makes customers feel pampered? You get the idea.

SOLAR CELEBRATION PARTY BEST PRACTICES

The most classic way to celebrate a sale is through a celebration party, in which the solar company provides some type of catered event at the homeowner's new solar house and is also present to educate those attending about going solar.

That's technique is nothing new. The solar guerrilla marketing idea here is to make this celebration unique, as mentioned above, not *just* a by-the-numbers pizza party or BBQ. Remember that the main goal here is to delight your customers and their friends so that they talk about you to their friends and enthusiastically refer you.

However you decide to celebrate, keep in mind these best practices for organizing the event.

1. **Assign a party manager within your organization**. This is someone who will be responsible for setting up each party and making sure a company rep will be there to attend the event.
2. **Ideally the company rep who attends should be the sales person.** Clearly, customers want to celebrate their install with someone who's gained their trust, not just a lead generator they've just met for the first time. That being said, if you have a natural host or hostess within your organization, that may be more efficient and effective, especially if the sales person is not very sociable with strangers. That would be odd for a sales person, but not unprecedented.
3. **Of course, bring brochures and other marketing materials.** Ask the homeowner if it's okay to bring brochures, CDs, jump drives, photos, or any other marketing materials that can be given to party goers who are interested. Tell them that you'll leave them on a small table by the door.
4. **If necessary, bring a small table or stand that you can set up by the door.** A music stand works great for this. They're both portable and thin, able to be placed

45

discreetly by the door. If a customer already has a side table or other natural area for your marketing materials, ask the homeowner if you can set the materials there.

5. **Bring a laptop with a PowerPoint presentation or a video.** This PowerPoint or video should consist of your basic solar FAQs, especially regarding costs, financing, and time factors. Keep it under 10 minutes. If the homeowners are okay with you making a presentation, then set a time in advance to have that presentation and a Q&A. If they don't want a presentation, then ask if you can be running a loop of the PowerPoint and/or the video on a laptop or iPad during the party.

6. **Be part of the clean up crew.** Don't just present, party, and leave. Some people may not want you to stay until the party's over, but at least offer to help clean up.

7. **Remember to bring plenty of business cards.** Obvious, but something for your checklist.

8. **Be on your best business casual behavior.** Once again, this should be obvious, but for the record, your rep should be a professional 100% of the time he or she is there. Male or female, wear a company polo shirt or long sleeve shirt with a branded logo, signifying who they are and why they're there. Don't drink at all, or one max, and of course, don't make passes at the hosts or the guests. If somebody really wants to get in touch with a rep socially, they can always pick up a card and call later. The goal for your attending rep is business, not social.

9. **Set a promotional expense limit.** Only you know what your promotional budget limitations are. As long as you tell your customer up front that you'll be offering a party that can typically accommodate X many people, your celebration should still be welcome. If you want to make a ceiling dollar amount, I wouldn't advise it. Some customers may be turned off by naming a monetary amount.

10. **Send a personal follow-up email the next day.** Be sure to connect with your customer the next day and say that you hoped they enjoyed the party. Make it personal, not boilerplate or automatic.

Solar Party Suggestions

Once again, you may have some unique solar party ideas for your area and company. Use them! If not, I hope these creative solar party ideas will inspire you—as well as your customers.

1) Celebrate with a solar wine tasting or champagne tasting party. Establish a relationship with a solar powered vineyard or local wine story. Not only can you educate about wine and champagne and solar, but you may also receive discounts and shared promotional expenses.

2) Celebrate with a solar beer tasting party. Same idea as the wine tasting party. You might also try a Beer and Sausage party.

3) Celebrate with a sunshine theme. Bring sunflowers, dress in yellow branded shirts, get sunflower table clothes, and party hats. Here, you can bring whatever catered food you want. The focus here is on the decorations more than the kind of food.

4) Celebrate with a Taco, Burrito, or Fajita Party. Partner with a local Mexican eatery who can cater a small party with Tacos, Burritos, or Fajitas, etc. Here the focus is on a popular food item. Once again, you may also get a discount from this local eatery for being a steady customer.

5) Celebrate with a donut/cupcake/ice cream/dessert tasting party. Have a local bakery that people love in town? It's always fun to celebrate with some desserts.

6) Celebrate with a private cooking class. Cooking classes are very social events, and cooking educators love to give classes. Find one who's open to have a steady stream of parties and work out a sun themed menu. For example, cooking with sun dried food, such as raisins, apricots, sun dried tomatoes, etc.

7) Celebrate with a private magic themed party. With solar, your high electric bills magically disappear! Find a local magician who can give a great in-home performance, and work out some solar themed magic tricks, if the magician knows of any or can adapt a trick or two with an electric bill or with a solar panel.

8) Celebrate with a solar independence party. It's always the Fourth of July with your company! During the warmer months, cater a hot dog and hamburger/grill party and celebrate your customer being more independent from their utility and dirty fossil fuels. It's also a great opportunity to explain net metering to guests, and the financial advantages of being just a little independent, as opposed to being off-grid.

Solar Gift Celebration Best Practices

Solar gift celebrations are basically "Thank You Gifts" for customers trusting you to install a solar system on what's typically their largest retirement investment.

- Whatever you give to celebrate the sale, try to make it something that is either useful or entertaining or both; also, if possible, lasting. A bouquet of helium balloons may be a fun gift, but once they've popped, the fun and the connection to your solar company is over. Perhaps a plant or planting an actual sapling tree would be a more lasting and memorable symbol.

- As with solar parties, consider your demographic and have a selection of celebration gifts for different demographics.

- If possible, **make these gifts a surprise** bonus benefit right after the installation is turned online and fully net metered/grid connected. Doing so is not only thrilling, but will make your company even more memorable.

- Include business cards and information about any paid referral programs you have. However, they should know about these paid referral programs before hand.

- Be sure to connect with your customer after the gift is delivered with an email, and say that you hoped they enjoyed it. Make it personal, not boilerplate or automatic. If they didn't receive it, find out why, and fix it.

Solar Gift Suggestions

1) Celebrate with a relaxing massage for two. After the installation, have the sales person present a gift certificate for two massages. The theme to communicate here is that they can relax with your company and with going solar.

2) Celebrate with solar wine or champagne. After interconnection, have the sales person present a bottle of wine and the gift of two champagne or wine glasses. Be sure to explain how the wine or champagne is solar powered. The same can be done with beers and beer glasses.

3) Celebrate with the movies. An ideal gift for families is a night out at the movies. After interconnection, give away 4 tickets to a local movie theater. Explain that they can now relax and enjoy their new solar system.

4) Celebrate with something that is delivered monthly. There are many Internet delivery services that will deliver a monthly selection of something. Wine, chocolate, fruit, flowers, bagels, coffee, etc. Every month for a year, your customer will be reminded of your solar brand and their new solar system.

5) Celebrate with a solar device or toy. After the interconnection and final hand off, give the family a wrapped gift of a solar toy or gadget. There are many portable solar powered toys, iPhone chargers, and external solar charging gadgets today. Every time your customers see or use that gift, they will think of you and their larger solar powered possession—their home.

6) Celebrate with a pair of his and her sunglasses. This is an obvious reference to solar, and if they're a decent pair of sunglasses, your clients will remember you every time they put them on.

7) Celebrate with a pair of branded coffee mugs and coffee or tea. Hopefully, your sales person will know whether you have coffee drinkers or tea drinkers or both. If they drink these beverages regularly, they'll be remembering your solar company with every cup of Joe.

8) Celebrate with a gift certificate to a local eatery. Is there a go-to local eatery that everyone in town loves? Make a deal with those guys and give clients dinner for two there. Include two glasses of solar power wine or beer, if the restaurant has them.

Tax Deductibility Considerations

Consult with your tax advisor about how your solar parties or promotional gifts are deductible. From my non-expert reading of the tax code, events that are open to the general public are 100% deductible and not considered meals and entertainment, but part of advertising and promotion.

"Gifts" may be deductible up to $25. However, some tax advisors may consider your gift, depending on what it is, an advertising and promotion expense. Once again, PLEASE CONSULT WITH YOUR TAX ADVISOR about how these solar celebration parties and/or gifts are a deductible business expense.

SUMMARY

- Celebrating every sale is a proactive and memorable way to generate brand loyalty and leads.
- Be creative with how you celebrate each solar sale. Make the effort to be unique to your company and your area.
- Be sure to include brochures, check-in emails, and follow-ups after your sale celebration. The more you can connect with customers in a positive way, the more they will remember you and your terrific customer service, as well as their solar system.

Select Previously Published Solar Guerrilla Marketing Posts

The following pages contain creative solar guerrilla marketing ideas that were previously published on my UnThink Solar blog on **www.RenewableEnergyWorld.com**.

I'd like to thank the editors and publishers of *Renewable Energy World* for hosting and publishing all of my Solar Fred rants, raves, ramblings, and calls for solar advocacy action.

A Solar Fred/"Spinal Tap" Solar Guerrilla Marketing Tip: 11

What if your installer's warranty went beyond the standard 10 years?

Perhaps the funniest moment of the classic mockumentary/rockumentary film THIS IS SPINAL TAP is when Nigel Tufnel, one of the band members, shows off the band's amplifier. He explains how his amp is "special" because it goes to 11. If you don't remember that scene, here it is again.

Watch it on YouTube at: **http://youtu.be/EbVKWCpNFhY**

Let's forget for a moment that Nigel has banged his head on one too many guitars. Instead, let's focus on his feeling special because he owns an amp that goes beyond the standard "10" amp.

Bringing that concept back to solar marketing, most residential solar installers guarantee their workmanship for 10 years, and as we know, good solar system installs are pretty stable and last far beyond 10 years.

Consequently, if you're looking for a way to stand out from the local competition, offering an 11 or 12-year workmanship warranty is a great solar guerrilla marketing method for many reasons.

In fact, in honor of SPINAL TAP, I'll give you 11 solar marketing reasons why an 11 or 12 year warranty is better than 10.

1. As mentioned above, it makes your brand stand out from the competition.

2. You give more to the customer with minimal risk, and unlike THIS IS SPINAL TAP, your 11-year promise is real, right?

3. During your sales consultation, it builds trust with the prospect who's concerned about putting holes in the roof.

4. It cements a longer (1 to 2 year) business relationship that allows you to receive more referrals.

5. It may help to close the sale if the customer has received competitively priced (or lower) quotes.

6. Can't think of anything to blog about? Write a post about your free extended warranty.

7. It's a great benefit to use in any print or direct mail advertising.

8. It can be highlighted on the front of your brochures.

9. It's a subject that you can address in a YouTube video, showing all the reasons why you're so confident that your service will last 11 years and beyond.

10. It can be highlighted on your website, instantly giving website visitors an extra level of trust. "Service You

Can Trust with One of the Longest Warranties in the Solar Business."

11. It's a gut-check for your skills/company. If you really need to stop at 10 years, ask yourself why. What are you afraid of? Whatever it is, address that fear with training, people, or using better products.

Of course, solar leasing companies have a parts and service contract warranty that typically lasts 15 to 20 years — including the inverter, BOS, etc. So, that's yet another reason (#12) why offering a longer-than-standard workmanship warranty (plus highlighting the long-term ownership savings), can help your solar company compete with the leasing companies.

So, perhaps Nigel had something there: An 11-year workmanship warranty is a solar guerrilla marketing way… to UnThink Solar. Next week, we'll discuss the solar marketing answer to the ultimate question of life, the universe, and everything. (You must be a *Hitch Hiker's Guide to the Galaxy* to get this joke.)

Drip, Drip, Drip: How to Effortlessly Educate Solar Customers with a Few Mouse Clicks

Here's an automated way to share solar's FAQs and cement brand loyalty.

One of the responses I received from a solar marketing survey that I did was from an installer who said that they didn't have enough time to educate all of their customers. That's understandable if you're a small solar shop with a few sales people. One solution is what's called an email marketing "drip campaign."

An email marketing drip campaign is basically a series of automated emails. With today's email marketing providers, you can set up a series of 1 to 10 solar education emails that can be sent to prospects who have voluntarily given you their email information.

So, say you get a call or email address from a residential solar prospect who is requesting more information about your solar services. A drip campaign may cut down the time spent on those calls or eliminate them entirely.

Here's how.

Step One: Get an email provider. First, you're going to need an email provider that has some type of automated response campaign. I personally like Mail Chimp (http://ow.ly/k5EVi), but there are many other programs out there, such as AWeber or Constant Contact. It will cost you $10 to $15 per month to do automated drip campaigns, but it will be worth it as a time saver.

Step Two: Set up a template and email campaign. One of the great things about email marketing programs today is that they're very cut and paste. You don't have to be a website designer to create a very nice looking email. Just pick a pre-designed template, upload your logo, add content, and you're good to go. Here's a great video I found that explains the mechanics of a drip campaign in Mail Chimp. See the video at **http://youtu.be/yQdmisKlnJl**

Go through those steps or the equivalent process with other email programs. In terms of frequency, you could set it up daily, every two days, or weekly. Try daily first, and if you get too many unsubscribes after the campaign starts, change the frequency to two days, or perhaps weekly. That being said, if you have good content, chances are that prospects will stay with you for the long haul--or perhaps contact you sooner for an official quote.

Step Four: Add solar drip campaign content. The above video shows you how to mechanically set up a drip campaign, but what content are you going to use for each "drip?" Think about it in terms of Frequently Asked Questions, as well as an introduction to your company. With that in mind, here are Solar Fred's suggested topics for 8 drips:

Drip #1: Thank you for contacting us. This first introductory email will be sent immediately after the person is manually added to the list from your brief initial contact over the phone, or automatically generated through signing up through your website. (Your email provider can provide a direct link to a sign-up form that triggers the drip campaign.) In this first email, describe your company in a sentence or two and explain that over the next few days/weeks, that they'll receive x number of emails that answer their most frequently asked questions about going solar.

- Include the titles of the suggested drip topics listed below.

- Also include a statement that if they'd rather get a free solar quote, they can email you at freequotes@yoursolarcompany.com. On the one hand, this defeats the purpose of automated solar education. On the other hand, if the person clicks that email address, they're more serious about making a purchase decision.

- Alternatively, you could provide a link to a PDF download of the same information contained in the entire drip campaign, but I think that can be overwhelming to new prospects. Instead, a full drip campaign gives them the same useful information in little chunks.

Drip #2: How much does solar cost? It's the top FAQ of any new customer. Briefly explain why the cost is different for every home and perhaps give a range of what it might cost. Also be sure to mention that there are different ways to finance solar, including solar leases/PPAs, but add that they'll receive more information about that in the next email. End the email with a call-to-action, such as "Ready to get a quote? Contact us for an appointment," and tell them about the upcoming topic that will be delivered next.

Drip #3: Available Solar Rebates and Incentives. Briefly explain all of your local rebates and incentives, as well as the Federal ITC and net metering. Once again, end the email with a call-to-action and include a mention of the next topic.

Drip #4: The Different Ways to Finance Solar. Here's where you explain about any financing options you have, including home equity loans or any national solar financing options. Once again, end the email with the same call-to-action, and prime them for the next topic.

Drip #5: Is My Home Right for Solar? Some might argue that this should be Drip #2, but cost is the consumer's number one question, so don't keep them waiting. But if you want to make this Drip #2, be my guest. In any case, include roof age, shading, roof area, owning the home, etc. If you want to reinforce credit requirements for solar leases/PPAs, do that too. As before, end with same call-to-action and mention of the next drip.

Drip #6: How does solar work? Briefly explain how solar works. Email programs allow you to use images too, so be sure to include helpful diagrams. Include the call-to-action at the bottom and set up the next drip:

Drip #7: How to choose a solar installer. Here's where you're going to mention all of the reasons why you're a quality solar installer. Frame it as "Qualified solar installers should have x, y, and z," and show that you have x, y, z, and more. That is, you're licensed with the state, you have a great BBB rating, you're NABCEP certified, have been in the business x number of years, and have performed x number of installs, etc. So, whatever solar authority you have, put it in. As always, end with the call-to-action and a mention of the (possibly) final drip:

Drip #8: What our customers are saying about us. In this potentially final drip campaign, add your verified customer testimonials. DO NOT MAKE THEM UP. If you don't have truly satisfied solar customers who are willing to take a call and recommend you, then you have more serious problems with your business than customer education. So, ask these happy customers if you can include their testimonial and their email address as a reference. Include a nice picture of the installation next to the quote, but don't give out their physical address. Make one last call-to-action to set up a time for a quote. If this is the last drip, say that and urge them to contact you for any further questions.

If you want to add more drips, go for it, but keep in mind that people have short attention spans today. So, make sure all of your drips are brief, and if possible, include helpful images.

The beauty of this type of solar education is that it really is effortless once you've set up your program and added the content of each drip. Of course, you may need to edit/refresh content or frequency. Nevertheless, it's a low cost/minimal-labor method to build trust and authority....and to UnThink Solar.

Puppy Dogs and Solar Quotes: An Example of Cause Marketing

Can genuinely supporting animal shelters also help drive solar leads?

Getting a free solar quote is probably the best way to open the door to a residential sale. People can finally see the numbers and get past their fear of solar being too expensive. The problem is that it can take a lot of time and marketing dollars to get people to get that quote. Wouldn't it be great if people were actually enthusiastic about getting a solar quote?

Well, that's possible, but for that to happen, you must first support something that groups of people are already energized to support.

I'm talking about what's commonly called "cause marketing." Instead of enticing prospects with a contest drawing for a free iPad with every quote, what if you sponsored a cause that you truly believed in and gave a donation for every quote? This strategy can be very powerful, because it can inspire people to go beyond their own self-interests when they believe in the same cause that you support.

For example, what if you sponsored a dog pound or shelter? Dogs and cats may not have anything to do with solar, but if you're genuinely interested in helping these animals, that passion can spread and touch your solar business in a positive way.

The idea is to utilize the cause's existing support for these animals. So, for this example:

- Contact the shelter manager and offer to donate, say, $10 dollars for every solar quote referral and $100 more dollars for every sale from a referral.
- Take photos of the dogs or cats with solar panels or playing in front of a solar home.
- If people in your company have pets that were rescues, include those stories and photos in the campaign materials.
- Sponsor an adoption day and volunteer to help set up. Bring a solar panel connected to a battery and a fan, or perhaps a coffee pot, or something else useful for the event.
- Using these images/videos and stories from these events, the shelter fundraising team can then work with you to send an e-blast, letter, and/or a Facebook post to their supporters — many of whom have homes — and urge them to get a free solar quote to support the shelter — and to learn if solar is a good fit. Specify that supporters must be homeowners to qualify.
- Regardless of how many quotes come in, provide a minimum and a maximum amount of money that you'll donate.

If you've done all of the above respectfully and genuinely, dog and cat lovers will share this donation information with other local dog and cat lovers. People may already be supportive of solar, but now they will be inspired to take the time to finally get the solar quote and perhaps be pleasantly surprised at how affordable it is. They may then tell their friends not only about the animal shelter donation opportunity, but also about their great solar quote experience. You'll also receive brand awareness and be associated with a worthy cause.

Not a dog or cat lover? No problem. What other causes do you support? A museum? Theater Company? Zoo? Church? All of these causes and more can help you get the word out about solar to their networks. But first, commit your time and marketing dollars to helping them.

Cause marketing is just another way… to UnThink Solar.

12 Things You Can Do in One Day to Market Solar

You can always do at least one thing to market solar in a day. Here are 12 suggestions. Choose one or all 12.

This Solar Fred marketing post is inspired by a recent Seth Godin blog post (**http://ow.ly/IZ787**). I really think Godin captures the spirit of how small but consistent efforts can make a difference — without having a million dollar ad budget.

Below is my adaptation of Godin's one-day-effort bullet points for solar marketing and innovation. Each of these 12 ideas can be inexpensively accomplished in a single day for little or no money. Pick one a day.

Repeat as necessary:

1) Send a handwritten and personal thank you note to a solar customer. These days, we rarely receive a handwritten note from anyone. Everything's so automated.

And yet, whether you're in the solar B to B world, or the residential world, that's precisely why you'd make a hell of an impression on a new solar customer by being as personal as possible. It doesn't have to be handwritten, but it does have to be specific and sincere. If written well, it will inspire referrals and pay handsomely for your time and thoughtfulness. Every sales person should do this.

2) Write a blog post about how someone is using your product or service. Blogging is the anchor of any solar social media campaign today. Many people ask me about what to write about. Writing about how people are using your solar service, whether for home computers or a dairy farm, is a great way to show new customers that solar is mainstream. Plus, it tells a story, and everyone loves a story—or a solar case study.

3) Research and post a short article about how something in your industry works. What's great here is that solar marketing isn't always about your company. People treat marketing articles like they're press releases, and they're not. If something in this solar industry is working or is innovative, applaud it. Congratulate that innovation and share it with your audience. Yes, they will get credit, but you will get credit for recognizing trends and innovation. If it's not perfect, share what could be improved. Be a thought leader, not just a reporter.

4) Introduce one colleague to another in a significant way that benefits both of them. Sometimes you may recognize when a colleague is missing some connection and needs help. Help them. Your generosity and thoughtfulness in this one day can lead to help and resources for you down the road. If not, you're still doing something good for your company, and you care about your solar company, right? If not, perhaps it's time to rethink where you're working and how you want to grow your career.

5) Read the first three chapters of a business or other how-to book. I take time to read at least one chapter of a marketing book every day. If not a marketing book, then at least a marketing blog post from a respected source like Seth Godin. Solar is still a tough sell, and so we must inspire ourselves with examples and advice from outside experts. Takes 10 or 15 minutes, and who knows what solar campaigns will come out of it.

6) Record a video that teaches your customers how to do something. Solar is perfect for how-to videos, and the great thing about solar products and installation is that you can easily shoot a video without interrupting work, and then use voice over to explain what's going on. We're in a visual world today. The more video how-to's you have, the more you will have authority and inspire trust with prospects.

7) Teach at least one of your employees a new skill. There are so many new people to solar, and they all do need some advice and tips from veterans. Once again, do this with an open "how can I help" attitude, not a "what's in it for me." What's in it for you is growing your company and building your internal business community.

8) Go for a 10-minute walk and come back with at least five written ideas on how to improve what you offer the world. I'm just going to echo this. It really does help, and I often come back with more than 5 ideas. Be sure to have a digital recorder or memo app on your cell phone to get it all down. You may throw away 4 ideas, but that 5th one may be the breakthrough idea sets you apart from other solar products or services.

9) Change something on your website and record how it changes interactions. It's good to surprise return visitors to your website. Change a photo, create a new page. Make people look twice at what they thought was familiar. The goal is to increase curiosity and make website visitors interact/email/comment more. Experiment. See what happens. If it flops, you can always change it back.

10) Help a non-profit in a significant way (make a fundraising call, do outreach). I'm going to do that right now. Vote Solar is having its annual Equinox fundraiser on March 18th in San Francisco. I not only spent the $100 for the ticket, but I'm flying up from Los Angeles and staying at a hotel. It's a great party and great cause for a great organization that's working for all of us. Hope to see you there.

11) Write or substantially edit a Wikipedia article. I've never done this before, but I wonder how many mistakes are in Wikipedia about solar power or inverters, or Solyndra. Whatever sector you're in, review the page that applies. If there isn't a page that refers to your kind of product, then you, my marketing friend, have a huge opportunity to educate the Wikipedia public about your particular solar product or service. Don't be promotional, because the Wikipedia editor elves will delete it. Just provide factual, useful information.

12) Find out something you didn't know about one of your employees or customers or co-workers. Once again, the point here is to build an internal solar think tank and to build a company community with strong, mutual, solar goals. So reach out. Have a monthly pizza party and request that everyone reveal something that people don't know about him or her. At best, the answers may inspire a new solar marketing campaign or perhaps inspire a product innovation. At worst, a cheap lunch.

Those are just 12 more ways... to UnThink Solar.

When Will a Solar Company Guerrilla Market Like this Bank?

Flash mob guerrilla marketing that has received over 10,000,000 views on YouTube.

Here's another example of a non-solar company taking a small guerrilla marketing risk and getting a 9-million viewer pay-off on YouTube. I repeat: Nine million people watching a five minute video made by an international bank that few Americans have ever heard of…until now.

Watch it on YouTube: **http://youtu.be/GBaHPND2QJg**

Industry pros often tell me that it's the installers that choose solar PV panels, and I agree that's the current reality for residential, commercial, and utility sales. But that doesn't have to be the case in the residential and small commercial sector if more solar PV companies (and installers) did something like the above every once in a while.

Wouldn't it be great for customers to *know* the name of a PV brand beyond Solyndra because they saw — and were inspired by — a creative solar guerrilla marketing video they saw on You-tube or Facebook? One more time: Over 9 million views!

So, why is this video so successful and how can it apply to solar companies?

1. It's visual. They chose a great spot in a public square, placed a camera in front of an odd looking, talented guy with a cello, added a cute kid putting a euro into the hat, and that was all the hook needed to keep us watching.

2. It's mysterious. Quickly, the single cello player is joined by more musicians, who are equally talented. What's happening?

3. There's a payoff. The mystery unfolds, and we see the entire spectacle play out, literally to a crescendo.

4. It's universal. Perhaps you don't know this is Beethoven, but you've probably heard that song once in your life. Even if you haven't, there's a reason why orchestras have been playing it for 200 years. It's a great song. ("Ode to Joy.")

5. It's positive. Sure, sex and violence sells, but only in newspapers, television, and video games. For brands that sell solar panels or installation, you want to be associated with something positive, and giving a free, spontaneous, concert on a public square is certainly positive. Note the many genuine images of children enjoying the music as much as the adults.

6. It involved the community. The Bank didn't just hire any concert orchestra from any city. They were celebrating their 130th anniversary in their hometown, so they looked to their own hometown orchestra and professional choir.

What was the point? The marketing point was brand awareness, but it's more than that. It's also cementing customer relationships, especially in the bank's home office city.

What does a classical concert flash mob have to do with an international commercial bank? (Banco Sabadell). As noted on You-tube under the video, this was a part of their 130th anniversary celebration. Their slogan is "We are Sabadell," and this flash mob reflects their boldness, their creativity, as well as their pride in being part of the city and its citizens.

Could a solar company do something like this? Yes, but it's too late now. It's been done and viewed by 9 million people around the world. However, there are many ways to do a flash mob, and if you're a solar company, create something related to solar. And by that I mean:

- Build something in public or for the public with solar panels.
- Power that something. If you want to make a storage point, have a battery to power it at night too.
- Make that something move. Solar energy may be quiet and still (on the surface), but guerrilla marketing isn't.
- Make it have sound, whether that means sound effects or music. (But don't just hire a band and power guitar amplifiers with solar panels. That would be obvious and boring.)
- Involve the public. Allow kids and people to push a button, pull a lever, whatever. Of course, be sure it's safe for onlookers to participate. If you have any doubt, keep it to your crew.
- Make it visually educational about solar, of course, but feel free to make other points too.
- Make the concept reflective of your brand. Even if you don't want this to reflect your brand, it will. So, might as well shape it. It's another reason why you should stay positive and family friendly.

- Whatever happens, whatever you do, videotape it with several cameras. That will allow for better editing choices later.
- Be safe and be legal.

This isn't rocket science, but it is solar science and there is some risk of falling flat. But wouldn't that risk be worth it? 9 million views…UnThink Solar.

Another Great Example of (Non-Solar) Guerrilla Marketing

Whatever you do, don't push that mysterious button in the middle of the street. But you will...and then what happens?

I love finding guerrilla marketing examples on YouTube. Unfortunately, I rarely find ones that are related to solar. Bummer. Nevertheless, it's always great to get inspired, and below is a wonderful model that you might be able to adapt...if you're willing to push that button.

Watch this video first, and then we'll discuss why it has nearly 80,000 views on YouTube, as of this writing, and we'll also go into how this structure might be adapted for solar marketing purposes.

Watch it on YouTube: **http://youtu.be/9OIJRMqYAA0**

So, why does this work? Let me count the ways:

1. It's fun. There's a smile on your face when you're watching this. You're saying, "Wow, this is so crazy, so cool, so amazing, so..." etc.

2. It's mysterious. Big button, middle of the street, and a sign that says "Push this button for drama." We're naturally curious humans, and we also like to be challenged. Sure, there's a little danger there, but that sign is so odd. Someone is going to have the courage to push that button.

3. It's surprising. If people pushed that button, and someone just shook their hand, that wouldn't go viral. What makes this work so well is that it exceeds our curious expectations. Not one, but many dramatic scenarios happen after pushing that button. The marketers here did not disappoint our expectations for "drama." Which brings me to...

4. It had a purpose. This was a crazy kind of stunt, but by the end you see that there was actually a method to this madness. The stunt of perpetual "drama" was related to TNT, an international cable channel that offers dramatic films and television. Therefore, these stunts were designed to mimic the dramatic scenarios that you might see every night on TNT. It wasn't push the button and see 20 clowns coming out of a Mini Cooper. That would be unrelated. Instead, you saw "drama" that hammered the point that TNT is the channel you want to be watching when you're in the mood for... action and drama.

5. It was video taped. Guerrilla marketing isn't going to be cost effective if only the attending audience views the stunt. This puppy was filmed so that others could enjoy it on websites, media, and social media, such as blogs, Twitter, Facebook, and of course, YouTube. So, whatever you come up with, film it to make it last.

How to apply this lesson for solar guerrilla marketing:

Essentially, this is "Pandora's Box" guerrilla marketing. You need to make a sign that points to an object and dares the reader to do some action.

With that in mind, choose a public place. If it's too mysterious (like the TNT example above), alert the police or internal security guards and let them know what's going to happen. A box, trunk, or button in the middle of the street or park plaza could be seen as a threat, so if anyone calls authorities, they'll tell them what's going on. Naturally, obey all local laws. If you need a permit to do street theater, get one.

It's up to your company's creative solar engineers to figure out the Rube Goldberg solar-related event that's going to happen when you hit that button, or lift up that box, or uncover those panels. Could you demo a small solar tracking system? Perhaps an AC/micro-inverter solar panel?

Huzzah. Go for it. But be sure to structure a beginning, middle, and end to your plan. That is, once the panels point towards the sun or are uncovered, then what happens? What's that solar panel juice going to turn on and energize? An over sized radio with a real rock band inside? An air conditioner? Huge fan on a hot summer day? A fridge with free, solar-cooled soda?

And what happens when the panels are covered abruptly or a cloud goes by? That is, what further unexpected thing will happen when the solar power disappears, either naturally, or by pushing another button?

Finally, what's your ending? Like the banner that comes down at the end of the above video, build in a fun way to identify your brand—and your point.

Just another fun guerrilla marketing way…to UnThink Solar.

FAQs about Solar Fred

How did you come up with "Solar Fred" as your solar pen name? Your first name is "Tor."

Really short answer: "Solar Tor" didn't sound right. Too weird for American audiences. As for choosing "Fred," I had a running joke (not X rated) with my ex-wife about the name "Fred." If we ever meet, ask me about it.

What does "UnThink Solar" mean?

It means to step back and rethink your usual solar marketing strategy. It's as if someone said to you "Think outside of the box." The box, in this case, is boring solar marketing and communications. In fact, I'd like to think that truly creative solar marketing has no box. You start with pure imagination. I hope my ideas inspire readers to boldly reach out to customers, and I want to encourage them to not only stand out, but also to educate customers. That's the gist of UnThink Solar.

Where are you based?

Technically, the San Francisco Bay Area, but since this is the age of the internet, we're really everywhere.

How can I contact you?

Info@solarfredmarketing.com. Or feel free to Tweet me at @SolarFred.

About Tor Valenza a.k.a. "Solar Fred"

Tor Valenza a.k.a. "Solar Fred" grew up in New York City and has been a solar advocate for over 25 years. He holds a B.A. from Middlebury College and is "fluent in solar," having taken numerous solar economics and technical courses at the Solar Living Institute and Solar Energy International.

After years as a successful writer and marketer in other fields, Valenza finally began putting his humor, communication skills, and solar passion to good use by founding UnThink Solar, a boutique communications and marketing agency devoted exclusively to the solar industry.

In addition to his solar marketing work, Valenza/Solar Fred is a leading solar industry blogger on *Renewable Energy World* where he's received over one million page views for his solar marketing advice and solar advocacy campaigns.

Valenza is also well known on Twitter, where over 7000 solar industry pros and advocates follow him for solar industry news, advocacy, and information.

www.ingramcontent.com/pod-product-compliance
Lightning Source LLC
Chambersburg PA
CBHW071613170526
45166CB00003B/1077